Radiesthesia III
Senses

ISBN: 9781787234475

RADIESTHESIA III

SENSES

**A wholistic approach to a story which determines
how life is conducted**

by

Christopher Freeland Ph.D

INTRODUCTION

While we will probably never know for sure where the truth lies. Is it even reasonable to believe that such a thing actually exists? How can we honestly expect to find a conclusion when we base our deductions on only a part of the whole picture? If life is an interconnected whole of which we are merely a part, would it not be reasonable to accept that there are some influences at work that are apparent to us, yet others which remain invisible but equally powerful in affecting what happens? The Sanskrit term for truth is so evocative of this state which we feel when we are in harmony with what is. The word is *sat*, etymologically derived from what we commonly express in English as *existence*.

Visible – invisible, material – spiritual, gross – subtle. We navigate through these strange waters of existence with little help from our fundamental education, but it is not because such matters are unknown to man. No, far from it, as anyone who spends time observing Nature will tell you.

This pamphlet aims at bridging this gap – one more – in the life puzzle by adding a little input to the obvious fact that we interface between the worlds of the manifest and intangible, thanks to our senses. They act as guides, faithful friends allowing access to those dimensions where we find the greater benefit.

We humans are good at breaking things, and a lot of effort goes into that practice. Differentiation, the practice of distinguishing between two or more things, by its very procedure does just that. So, in this effort at defining individual senses I am as guilty as everyone else with the simple difference that the thrust of my whole approach here is to integrate all the components, having applied my intuito-analytical ability, rather than leaving them separated. Even though those parts were never apart in actual fact.

The first question with regard to the subject of 'senses', could well be, how does one define a sense? It is evident that the so-called 'five senses' fall far short of the mark if we are speaking of the multiple capacities found in humans, animals and plants that allow us to make our way through life. So, throughout history, the scope has never been too well defined, and we are left to our own conclusions.

That being so, and although there is little likelihood of achieving a definitive answer, it struck me that it could be useful to have a better understanding of the interface between our apparent 'inside' and 'outside'. We, as humans, all experience a life cycle, consisting of waking, dreaming and sleeping, a mutually dependent and essential sequence that takes place in what appears to be a commonly shared environment. Is it not possible that our 'senses' are the essence of that interface? Acting on that possibility, I set about trying to make a nomenclature. Of course, with a difference (it wouldn't be fun otherwise) using radiesthesia. To my knowledge, no one has ever attempted such while putting the findings to a test of veracity – which I would strongly recommend you do too.

BACKGROUND

Some human senses have a seat, an apparent location in the body from where they operate, in that case we conveniently speak of a sense-organ. It would be quite easy to appreciate Aristotle's haste in deciding that we have five senses on that basis (apart from his ongoing tussle with Democritus), but if it is true that where he went to school – for twenty years – in Egypt, there was apparently (school of Kemit) a teaching whereby a human has 365 senses (unfortunately there seem to be no precise references in either the literature or epigraphic detail).

There are numerous body parts – pituitary, pineal, thymus, islets of Langerhans, etc. in the endocrine system, sensory neurons in the nerve ganglia, and so on – about which we have limited or no understanding. It would be a shame to neglect their potential function just because we do not have the approval of science as to their precise use. Especially if we are to break with the materialistic attitude which is the inevitable consequence of believing that we, as humans, have a mere five senses.

The notion of there being only five senses has been consolidated over time. Perhaps even a perverse manipulation to encourage the human to believe that he or she is apart from the rest of Nature, a being that does not share the same environment as all other creatures; for that is one of the consequences experienced in the last few thousand years.

We are told that we have lost the other senses, that is a myth. Like muscles which atrophy through lack of use, a sense could well suffer the same fate.

The pendulum, my tool of predilection, is used to find an answer for every question. I know no more than the next person but am convinced after fifteen years of intense training and constant practice that one can obtain answers to questions if carefully formulated, and that using any other form of kinesiological method. The key is that the question be as well prepared and unambiguous as possible.

Throughout the whole period of drafting this document, especially when the reasoning process was influenced by my own thinking, I used a double blind test by placing 'yes' and 'no' in two separate envelopes, shuffling them (or having someone else shuffle them), and then asking with the pendulum which envelope contains the correct answer.

To give an example, it seems logical to assume that the sense of thirst can have three potential categories which we are able to sense: excess, depletion, sufficiency. But is that the end to the matter? So, by asking the question "Is there any further dimension to the sense of thirst?", the answer is "No". "Have I omitted to ask something in this context?", again, "No". "Are these the only three categories to be considered as pertinent to the sense of thirst?" "Yes". Taking the two envelopes and placing them on the desk, "Are the answers provided correct?". On opening the envelope indicated, the questioning either comes to a close or needs to be reworked and started again.

The way the senses interpret the outside world is very important as that very fact determines 'inside' and 'outside', in other words the individual's reading of the information they personally receive on which they act, or not. It would also infer the very relative reality of the external world, a chimera of one's mental projections, or something else? That truth business again!

There are a number of notions introduced here regarding the nature of a 'sense' which may surprise – I hope not shock. So, you are very much encouraged to check and verify what I say, and please do not hesitate to make suggestions to me in person on thaitoff@gmail.com, or modify your copy to suit your understanding. Anyway, here goes in a modest effort to sort out one more conundrum.

1. Unicity

The foremost sense in theory, and for some in experience, could be said to be the feeling of being-at-one, atonement if you like. By nature, it is a continuity; has there ever been any doubt in your mind whether you exist? The doubt is one of form, not foundation, giving rise to duality – me and you, and her offspring, multiplicity – the world. That sensation of oneness can never be affected. In none of the three states humans experience, viz. waking, dreaming and deep sleep, is that gnosis absent. But we tend to pass it by because we do not affirm its existence, and most of our socio-philosophical systems teach us to look outside for some form of governing influence, rather than applying introspection which would soon teach us that there is no inside nor outside. Its most common name is Love. It is much easier to love your neighbour, as we are told is the right thing to do, when you know you ARE your neighbour!

2. Consciousness

We are clearly confusing consciousness with "understanding," "awareness," and "thinking". There is a bad case of mistaken identity, or, more simply, the wrong nomenclature, which needs to be straightened out right from the very start, or we risk lapsing into the bad habit developed over our own life time and repeated through history. All this modern talk of consciousness, as in "a crisis of consciousness," "a need to raise consciousness," "transformation of consciousness," and so on, is due to a sloppily or inexactly defined term.

You might well agree that the nature of consciousness is "light," the main source of brilliance illuminating the senses, so allowing intelligence, inspiration, perception, cognition, and revelation. A substantially different beast from the electromagnetic and infrared frequencies supposedly from the sun which enlighten and enable our waking state.

Consciousness in the *Oxford English Dictionary* is considered to have several synonyms, awareness and wakefulness being among them. *Wakefulness* says it all, but do you hear the essential nuance of context that that word contains? It reveals about one-third of the truth. Just because there is consciousness in the waking state, does that mean it is the only state that consciousness illuminates? Definitely not, but that conclusion renders "consciousness" one of the most poorly understood terms of our epoch, a typical example of how treacherous language is.

So can we redefine Consciousness?

First, with a capital *C*, because it is the *only* one there is.

A good understanding of the term Consciousness is provided by nondualistic Vedanta (the confidential teachings of the Vedas as exposed by Shankara, *kevala advaita Vedanta*). This is as good a time as any to expand on that teaching for those not familiar with this unique standpoint.

There is a common belief that Eastern philosophy is about "becoming" one with the All. That may be so, but from the Vedantic viewpoint it would be a grave misconception, because there can be no question of becoming for the simple reason that you ARE ALREADY

ONE WITH THE WHOLE. Due to ignorance or nescience of the true nature of yourself, however, you do not or choose not to see it. Ignorance is ignoring something once you know its nature, nescience is not knowing something; two very different beasts.

The very idea of trying to define Consciousness is one of those anomalies of human understanding. If we all have it and nothing can exist without it, how is it possible for a component part to define the whole? What's more, why is it necessary to limit it with words, as if it were something that could be measured? Such reasoning is radically opposed to its very nature.

Consciousness is limitless, one, omnipresent, all-embracing. In other words, a real whole shared by one and all, it cannot be differentiated by a category of species or protein (animal, vegetable, or mineral) or broken down into parts such as atoms, neutrons, or whatever; it defies classification whether divine, human, or mane, for the simple reason that it is the foundation and cause of all that exists. The downside of this is that Consciousness can be neither affirmed nor denied; reason can neither prove nor disprove it; it is not a subject of belief as a possibility, or rejection as an impossibility; and it is perfectly beyond the reach of mundane words.

The conviction of my own reality is intuition based; no one can dispute my Self. Even if they did, I would not be in the least affected—I know that I am. I employ the first person singular deliberately to emphasize the individual nature of this universal persuasion and would further argue that this holds true for every thing one encounters in this universe. However, if one were asked to prove the reality of anything other than one's own Self, one would insist on some very solid evidence and still likely remain sceptical.

Your own individual experience is called to witness here; you are not expected to adopt Hindu philosophical tenets, let alone adhere to what is written in the Upanishads now or in the future.

That we have lost the capacity/sense/knowhow to tune in to the world is in all likelihood the outcome of exaggerated egocentricity, perhaps the influence of materialism, or more specifically a belief in human superiority. It seems a reasonable supposition, in keeping with the individual intuitive experience, that we are in intimate connection not just with one another but with all that surrounds us, both animate and inanimate. But such a credo requires a quantum leap in understanding and a considerable humbling, because it means that you, as a human, are somehow on the same footing as the insect that just flew by, the unseen amoeba in your glass, or the distant mountain. It requires a progressive replacement of the ego—as the centre of your experiential subject-object world—as you gradually come to realize that in actual fact the ego never really existed as anything other than Consciousness.

I would maintain that this is the source of happiness for one and all, being in reasoned and sensually experienced balance with all in the environment.

The Nature of Consciousness

Duality and its immediate successor on the world stage, multiplicity, can be viewed as a consequence of the seeming appearance of differentiation, with truth disappearing like an ice cube in a glass of hot water, while the reality of unicity gives way to the usurper.

A brief explanation as to how Vedanta reaches this conclusion might be welcome for those not familiar with its methods. In passing, it is significant that there is no term in

Sanskrit for the concept of matter, which might indicate the importance given to the material by the designers of the system.

Consciousness is, come hell or high water, present in all the states known to humans; it is ALWAYS there. Perhaps it could be expressed as the unqualified principle of gnosis (subject/object-free knowing), unobstructed by a subject (knower) or object (known)—the intuition that you are aware, able to perceive. Even renowned Buddhist, Mathieu Ricard, says in his useful book, Happiness: A Guide to Developing Life's Most Important Skill, "That faculty, that simple open presence, is what we may call pure consciousness, because it exists even in the absence of mental constructs." Of course that is not pure, orthodox Buddhist teaching, because consciousness is momentary for the proponents of Siddhartha Gautama, but in essence very true.

In all likelihood, Consciousness is probably shared by everyone and everything around you, not just sentient beings. It is the intimate sensation that pervades all three states of waking, dreaming, and deep sleep. It has nothing whatsoever to do with the feelings caused by neurons or the nervous system, which, though of a material nature, need its total intimate support. Consciousness is omnipresent and all pervasive, requiring no input, although input cannot be perceived without it. Awareness is not Consciousness but cannot happen without Consciousness.

3. Awareness

Because of their intimacy, awareness is generally confused with consciousness. The definition of awareness requires an object to be aware of, therefore a *someone* to be aware of the *something*, and a context, as in not being aware due to the nonreceptivity of the senses?

Awareness of something is the result of its perception by one of the sense organs. One can well imagine that this happens due to the agency of conjunction of perception and the object, which for both probably occurs thanks to Consciousness and some form of resonance or tuning. As Consciousness remains a constant without any need for a perceiving subject, object perceived, or conjunction of the two, it is also the raison d'être of existence and no different from it. Similarly, reality depends on Consciousness for its being, not on the way people think or behave. That is not, however, what we are taught.

Most of us experience life as a continuity of existence. We know, as in gnosis (or "intuit" if you prefer), only one thing throughout our entire earthly existence—"I am." Even when you awaken after sleeping well, you are convinced that it was you who slept and that the state abandoned prior to sleep resumes on waking. Some thing was there to witness the experience. This knowledge is an enduring companion and a most convincing one; what's more, it is consistently encountered in all the three states. Thanks to the constancy of Consciousness, we have a thread to our existence, and it is the only unchanging constant. Now there is something that deserves a moniker for itself, and that is the sense given to it and will be used here.

This "I am" thread, experienced while awake and in dreams, is undeniable. If you look back over the past, even if there have been immense transformations in your life, changes of personality perhaps, careers, companions, there is consistently the notion of "I" that runs as the underlying associate. Ultimately, this "I" is all we have to work on in a quest for the exact nature of our elusive individuality. Much has been written elsewhere on this subject, so this is probably repetition for many, but it nevertheless remains a fact of life. The "I" is here to stay for the duration.

It would seem very normal that this "I," being nothing other than Consciousness—the unique component shared by one and all—can assume aspects such as what we refer to as the conscious, subconscious, unconscious, or superconscious component of the mind for the simple reason that we are not truly familiar with its real, all-encompassing nature, and we need to express the apparent changes as a comparison with matters that correspond to what we believe we experience, know, and, of course being human, think we control. Such terms are merely adjectival to the principal noun, Consciousness.

From the very outset of our human life, we are easily duped into believing this "I" to be something other than what it truly is. It starts out with the notion that "something is not quite right," a malaise that can assume huge proportions (even to a state of mental depression) if the call of the spiritual is denied. Refusal to listen to one's conscience, that inner voice that soon learns to keep quiet in view of the barrage of cerebral resistance opposing it, is a sign of the material gaining the upper hand. If one manages to maintain a balance of head and heart, there is a chance that much stress can be avoided. This state of object-less calm is harmony.

With the passage of time, the effects of education, peer pressure, simple habit, and other factors, we learn mistakenly to assume the reality of the physical body as being a thing of constancy, with an increasingly vague notion of what "I am," or what is this mind thing that is somehow attached to me, and an even vaguer idea that spirit is involved in the overall picture. Yet all the while, we *know* that there is something more to life than the material body.

Consciousness is this inner thing, this voice that is especially familiar in childhood, in dreams, moments of intuition, meditation, that suggests there is more than meets the eye in our world, so influenced and even fabricated by intellectual reasoning and logic?

Awareness, the first step in differentiation as Consciousness moves, so to speak, into realms where we exteriorise and lose the centre.

These three form the backdrop of what would seem to be the basis whereby we perceive, receive and interact with external and internal stimuli, but awareness is the interface between the information conveyed by the sense and what, if any, action is taken.

4. Right

Moral considerations apart, humans have a distinct sense of what is right – for themselves, others and their surroundings. Perhaps more often encountered as a child – the voice of conscience – until such time that this inner voice is extinguished.

5. Wrong

Much the same idea as above.

6. Existence

We all know we exist – I hope so at least. But it is this very powerful sense of being which leads to the sense of awareness, mentioned above that in turn provides a sense of satisfaction.

7. Satisfaction

There is nothing that is done in our human dimension that lacks this objective, which is very real, especially when not achieved. That is the driving force in any and every effort.

8. Nescience

An awareness that you do not know something which will either push you to find out and overcome the non-knowing or remain in blissful ignorance – deliberate turning a blind eye.

9. At ease or no tension

A sensation of comfort, the sensation of well-being, probably produced by the parasympathetic nervous modality.

10. Ill-at-ease or under tension

A sensation of discomfort, for whatever reason, caused by the sympathetic (fight-or-flight) nervous modality.

11. Communication

Interconnection could be a synonym for this ability. How many means of communicating with the outside world are there? They all probably use other senses or sense-organs to accomplish that task, such as eyes, speech, touch, ESP (it is extra because nobody bothered to write about how it actually works!), dream, telepathy, and so on. The bottom line is, nevertheless, that in an effort to re-establish the unicity of life, we inevitably revert to the harmony of Consciousness which is readily expressed through the immensely comfortable feeling when those lines of communication enabling connection with everything around us, are open.

12. Time

Not so commonly found in modern man as a sense per se, for the simple reason that time has become our governor rather than a collaborator which would be the case if man was to live closer to Nature. This is probably the most obscured sense, but the evidence is there that this sense is still with us. As for example when a person wakes just before or at the moment when the alarm clock is set; or knowing when to plant seeds because on is in sync with the moon.

13. Space

Here again, a sense that has been atrophied over the years due to non-application. It is experienced as a feeling of emptiness, the fact that there is nothing in your immediate surroundings that might be a threat to your well-being. It would perhaps be more correct to name it aether, for in the esoteric traditions, the aether has a faculty of connection and communication.

Motion

The body is aware of when it is in movement, at rest or not and determines the necessary actions as a consequence.

14. Upwards
15. Downwards
16. Backwards
17. Forwards
18. Left
19. Right
20. Diagonal
21. Fast
22. Slow
23. Stop

Balance
There is a balance mechanism, science would have it as being the inner ear, which provides a sense of being:

24. Upright
25. At an angle
26. Horizontal
27. Off balance
28. Balanced
29. Gravity
30. Acceleration

Magnetism
The affinity or aversion one feels towards the 'other', whether another person, situation, position, substance and so on, are probably governed at some level by magnetism, with its specific characteristics:

31. Attraction
32. Repulsion
33. Neutral

Temperature
34. Hot
35. Cold
36. Comfortable

Pressure
37. High
38. Low
39. Neutral

Thirst
40. Sufficient

41. Depletion
42. Excess

Hunger

Of all the mundane senses, this is the one which can systematically ensure good health in the modern age where over-eating has become the norm. Why is belching considered by almost all cultures as a sign of satiety? It is quite simply a signal from the stomach that the last bubble of air left vacant in the stomach is now on its way upwards, and NOTHING more is to be sent down.

43. Sufficient
44. Depletion
45. Excess

Danger

46. Threat
47. Relaxed
48. On guard

Dimension

By applying the Hindu concept, absent in all other philosophical systems, one acquires a substantial handle on life when considering the three states experienced by humans, rather than just the waking state. Although we only have a definite sense of which one we find ourselves in retrospect when we transit from one to the other, the experience remains very real in the duration. The doubt will always be there, however, are we dreaming a dream?

49. Waking state
50. Dream state
51. Deep sleep

Breath

We take an average 21,600 breaths in 24 hours, but the breath in and out of the lungs provides more than oxidation of the blood; breathing provides a different physiological function which could be said to have a related sense, especially when it is working properly:

52. Propulsive breath
53. Digestive or downwards breath
54. Distribution or nourishing breath
55. Upwards breath, responsible for sneezing hiccup, cough, dreaming and evacuation of the subtle body at the time of death
56. Breath of motion, responsible for joint movement
57. Suffocation

Taste

58. Sweet
59. Bitter

60. Astringent
61. Aromatic
62. Salty
63. Sour
64. Pungent
65. Umami
66. Maillard
67. Fat
68. Protein
69. Metallic
70. Danger (as in poison)
71. Spicy hot
72. Bland
73. Hot
74. Cold
75. Warm

Smell
76. Fragrant
77. Resinous
78. Fruity
79. Chemical
80. Minty
81. Sweet
82. Citric
83. Pungent
84. Decay
85. Durian
86. Popcorn (yes, the latest discovery!)
87. Humidity
88. Fear
89. Sexual or hormonal reaction

Touch
90. Hot
91. Cold
92. Pressure
93. Feeling as in sharing
94. Tickling
95. Itching
96. Irritant
97. Soothing
98. Rough
99. Wet
100.   Dry

101. Soft
102. Hard
103. Pain on the skin
104. Pain in the bones
105. Pain in an organ
106. Vibration
107. Contact
108. No contact
109. Change
110. No change

Sight

Probably the weakest of our senses, yet the one on which we seem to rely so intensely.

111. Light
112. Colour
113. Red and its stimulating effect on activity and effect on strength, with the longest wavelength at 625.740 nm.
114. Orange, less intense than red, 590-625 nm.
115. Yellow stimulates communication, optimism, creativity and memory at 565-590 nm.
116. Green soothes, at 500-565 nm, the colour of balance.
117. Blue calms and helps focus, 485-500 nm.
118. Indigo aids introversion, 450-485 nm.
119. Violet uplifts while calming mentation, 380-450 nm.
120. White covers about 400-780 nm allowing access to all frequencies.
121. Black has no frequency.
122. Brightness
123. Darkness
124. Conversion of light
125. Focus
126. Danger
127. Harmony
128. Vacant
129. Jarring
130. Blind spot
131. Master eye
132. Dream sight

Hearing

Direction of sound is calculated thanks to the lapse of time for sound to reach our ears, or rather that part of the brain which deals with sound. To the extent that animals with small heads are more sensitive to higher frequencies, we apparently, can only go as low as 12-20 Khz.

133. Harmony

134. Rhythm
135. Jarring
136. Blurred
137. Perceptive
138. Danger
139. Master ear
140. Direction
141. Density
142. Depth
143. Location

<u>Physical urges</u>
144. Defecation
145. Urination
146. Evacuation
147. Vomiting
148. Libido
149. Frigidity
150. Indifference
151. Revulsion
152. Fright
153. Flight
154. Approach of death
155. Conception

<u>Emotions</u>
The fact that a feeling arises in a physical being would indicate – in my thinking – that there is an attached purpose, or sense affecting the ego and subsequent behaviour. The word itself provides a clear insight to what happens: e = from outside, motion = influential movement. A shift in the outside environment (from solar flare to partner's mood, ...) results in an effect, that is what potentially impacts the individual.
156. Affection
157. Altruism
158. Amusement
159. Anger
160. Angst or dread
161. Annoyance
162. Anxiety for the outcome
163. Apathy
164. Arousal
165. Awe
166. Boldness
167. Boredom
168. Calm
169. Confidence

170. Confusion
171. Contempt
172. Contentment
173. Cowardice
174. Creativity
175. Curiosity
176. Depression
177. Desire
178. Despair
179. Disappointment
180. Disgust
181. Dissatisfaction
182. Distress
183. Ease
184. Ecstasy
185. Elation
186. Embarrassment
187. Empathy
188. Envy
189. Euphoria
190. Excitement
191. Fear
192. Fearlessness
193. Frustration
194. Gratitude
195. Grief
196. Guilt
197. Happiness
198. Hatred
199. Hope
200. Horror
201. Hostility
202. Humour
203. Hurt
204. Hysteria
205. Indecision
206. Indifference
207. Inferiority
208. Insecurity
209. Interest
210. Irritation
211. Jaded
212. Jealousy
213. Joy
214. Loathing

215. Loneliness
216. Love or affinity
217. Lust or carnal desire
218. Maternity
219. Melancholia or over-thinking
220. Misery
221. Panic
222. Passion
223. Paternity
224. Pity
225. Pleasure
226. Pride
227. Rage
228. Refusal
229. Regret
230. Rejection
231. Remorse
232. Respect
233. Sadness
234. Satisfaction
235. Self-image
236. Shame
237. Shock
238. Shyness
239. Sorrow
240. Suffering
241. Sufficiency
242. Superiority
243. Surprise
244. Sympathy
245. Terror
246. Weakness
247. Wonder
248. Worry

Action

249. Need to react
250. Need to remain still

Planets

There can be no doubt that the planets influence us and everything here on earth with their frequencies. Obviously, they form combinations of effects, affecting certain objects (people, plants) more than others, as a function of the aggregate frequency of the object. I consider them as formative forces of the human species due to their peculiar habit of influencing so much. This is where the accomplished astrologer demonstrates their sensitivity – for it is indeed a very intuitive affair – to compute these combined effects and provide a coherent reading. The inspiration here is that there is a definite impact from this relationship on the individual and, as a result, that affords planetary influence the status of a sense. I rest in the scope of traditional or classical astrology and do not include the recently discovered planets of Pluto, Neptune, Uranus, Chiron, Eris, etc.

251. Sun
252. Moon
253. Saturn
254. Jupiter
255. Venus
256. Mars
257. Mercury

Mental faculties

258. Mentation
259. Memory
260. Cerebral intellect or nerve capacity
261. Intuitive intellect
262. Egoity

Miscellaneous sensitivity

263. Radiesthetic faculty
264. Divination
265. Déja vu
266. Telepathy
267. Psychokinesis
268. Remote viewing
269. Premonition

Magnetic reception

It is astonishing to learn that science does not know how magnetism works. Yet we live in an environment that screams magnetism at every corner. This is the subject of the fourth volume in this series.

270. The magnetic faculty of the human

The elements

These are the physical aspect of the manifest equation, developed/developing over the millenia due to environmental conditions. As we have traces of them in our bodies, it is probable that they have an influence of some sort in the interface process. Apparently there are ninety elements that occur naturally in appreciable amounts, so I thought it reasonable enough to presume that older civilisations managed to find what we know outside of the laboratory. Consequently:

271. Actinium
272. Aluminium
273. Antimony

274. Argon
275. Arsenic
276. Astatine
277. Barium
278. Beryllium
279. Bismuth
280. Boron
281. Bromine
282. Cadmium
283. Calcium
284. Carbon
285. Cerium
286. Cesium
287. Chlorine
288. Chromium
289. Cobalt
290. Copper
291. Dysprosium
292. Erbium
293. Europium
294. Fluorine
295. Francium
296. Gadolinium
297. Gallium
298. Germanium
299. Gold
300. Hafnium
301. Helium
302. Hydrogen
303. Indium
304. Iodine
305. Iridium
306. Iron
307. Krypton
308. Lanthanum
309. Lead
310. Lithium
311. Lutetium
312. Magnesium
313. Manganese
314. Mercury
315. Molybdenum
316. Neodymium
317. Neon
318. Neptunium

319. Nickel
320. Niobium
321. Nitrogen
322. Nobelium
323. Osmium
324. Oxygen
325. Palladium
326. Phosphorus
327. Platinum
328. Plutonium
329. Polonium
330. Potassium
331. Promethium
332. Protactinium
333. Radium
334. Radon
335. Rhenium
336. Rhodium
337. Rubidium
338. Ruthenium
339. Samarium
340. Scandium
341. Selenium
342. Silicon
343. Silver
344. Sodium
345. Strontium
346. Sulfur
347. Tantalum
348. Tellurium
349. Terbium
350. Thorium
351. Tin
352. Titanium
353. Tungsten
354. Uranium
355. Vanadium
356. Xenon
357. Ytterbium
358. Yttrium
359. Zinc
360. Zirconium

The five traditional elements:
361. Water

Life is allowed here on earth thanks to water. It is not a coincidence that 75% of the earth's surface is covered by water, nor that 73% of the average human's weight is made up of water. Yet, we know so little about its real life-providing function. There is not the space here to elaborate further, so matter for another volume perhaps!

EPILOGUE

There will, in all likelihood, never be a final word on the subject of the Senses but as is so often the case, there can be no harm in trying to correlate what was apparently believed and held as being useful in the past, and modern research and development. This is my small contribution, to be taken in context and hopefully for others to expand on.

END

About the author

The basis of this text is founded essentially on a combination of practical experience, the study of theory from selected books, substantial personal practice and experimentation in a variety of methods and traditions. It all started in India in the early seventies when I studied Sanskrit so as to read the ancient texts in the original rather than relying on translation. In that phase, I had the opportunity to study with Swami Pranav Tirtha, a *dashnami sannyasin*, who initiated me into the Vedanta philosophy. Whilst with him I read, studied and assimilated the orthodox teachings of the Upanisads, the Brahma Sutra, Gita and multiple metaphysical and sundry texts of Hindu literature. I was ordained as a monk, with the name of Swami Chidananda Tirtha in May 1973. This period also furnished the occasion to study medicine with Dr Himatlal Trivedi, an Ayurvedic practitioner from Palitana, whom I accompanied in India and Africa in his practice amongst English and Gujerati-speaking patients. That involved study of the Hindu medical classics (Caraka Samhita, Sushruta Samhita, Ashtanga Hrdaya), with considerable practical experimentation of fasting and dietary regime on myself. Observation with Himatlal's guidance and explanation gave me a reasonable understanding of this medical art form.

Whilst living in France, I had the opportunity to study for a year with a French acupuncturist, who was persuaded to come out of retirement to teach Traditional Chinese Medicine (TCM) again. That grounding was then followed by many years of studying the Chinese classics: the Lingshu, the Huainanzi, the Suwen, along with in-depth reading of Soulié de Morant, Claude Larre and Elisabeth Rochat de la Vallée, in addition to extensive practice of moxibustion and acupressure. Whilst living in Chiang Mai, the opportunity arose to learn and practice a form of bio-energetics which involved a lot of practical moxibustion. The outcome of my studies of TCM affords a certain ease with this very complete approach to the human condition.

A vast amount of research, reading and experimentation with a very broad spectrum of subjects, traditions and cultures, combined with travelling and living among natives of other lands, along with my professional activity as a technical translator, specialized in nuclear and telecom technologies, for some twenty-five years in France have hopefully been turned to good advantage.

Study, practice and research into magnetism, laying on of hands, geomancy, radiesthesia and radionics add to my wholistic comprehension of life. A practice of organic farming combined with animal husbandry, special care for water and its supply have also led to my current understanding. I work more with the intention of clarifying what I think seems to be happening in the dimensions we evolve in, rather than any kind of dogmatic laying down of law. Since living in Ireland for the last three years, my focus has been specially on bio-magnetism, astrology and Egyptology, the latter being at the origin of this volume.

Whilst I sincerely believe my opinions to be correct because corroborated by the pendulum and experience, it would be a substantial error to think this is the last word because it concerns uniquely what falls within my own sensory (all three hundred and sixty-five) parameters. Too much is changing too fast for our perceptual ability to stay abreast of events, even were we able to comprehend and adapt. Perhaps it is not for us humans to determine how the ordered immensity of Nature works; that would be most presumptuous and dishonest, but it does seem to be worth trying to establish a mode of operation that might serve as a guide or possible reference fitting into our journey through this phenomenal existence.

I stand on the shoulders, hopefully in rectitude and fidelity to the thrust of the original argument of many researchers, practitioners and remarkable people in aligning these words on paper. The words will, as always, be symbols of the generosity of Nature as she carefully keeps everything in its structured place, although the human component must be the most unruly, hence the hard lessons we have to learn if we aspire to some other form of existence than the purely material.